STEP-BY-STEP
GUIDE FOR HVACS
INSTALLATIONS

Practical book for beginners on refrigerator and furnace repairs

Williams V. Dickson

Table of Contents

CHAPTER ONE..3

  INTRODUCTIONS.......................................3

CHAPTER TWO 9

  PARTICLES AND PARTS DEVICES 9

CHAPTER ONE

INTRODUCTIONS

PROCEDURES IN MOUNTING PARTED AC

Set and align locations

Do decide the specific area of the indoor and outside unit. You would rather not introduce the forced air system where the cooling will be insufficient or it will cool such a lot of that you feel like you are in the Cold Locale.

Check wall level

Ensure that the wall that will hold the climate control system's indoor unit is sufficient. Getting through the vibration of the unit ought to be capable. It is suggested that the indoor unit of the split AC has something like 11 - 15 cm of open space around its top and sides.

Current viable

This is required for appropriate wind current. Repressing wind current could make it harder to oust heat. This will drive your climate control systems working expenses up. It can possibly harm the blower also. In an 11 ft high room, for ideal cooling, a level of 6-8 ft over the ground is suggested. This will guarantee that the cool air spreads well around the room. It additionally guarantees that the climate control system isn't excessively bustling cooling the warm rooftop which just means higher power bill for you.

Use space

Open space is great for the outside unit of a split climate control system, maybe a terrace, in the balconies, back of side wall and furthermore on Pair chunks likewise so that it is completely safe to get to both for establishment and overhauling.

Open space truly intends that nothing remains to be impeded the stream or air. This will guarantee a great deal of opportunity for your blower and condenser to perform well.

Surface and allow vibrate

Do ensure that the outside unit isn't straightforwardly under daylight. In the event that a porch isn't accessible, an outside space with areas of strength for a can hold the heaviness of the open air unit without limiting wind stream, is likewise great. It is likewise vital to guarantee that the open air unit is on a level surface. Open air unit houses every one of the essential pieces of your climate control system including the blower and condenser.

Expand create

Outside units vibrate. In the event that the surface isn't level, consistent vibration, particularly during midyear months when the climate control system is utilized nearly relentless, could prompt harm of blower and condenser. A harmed blower and condenser are no decent.

Viable exhibit

This will expand your administration costs and in most pessimistic scenarios, may require the actual unit to be supplanted. Distance between both the units is vital to guarantee ideal and productive execution of your climate control system; the nearer the distance, the more viable the exhibition.

CHAPTER TWO

PARTICLES AND PARTS DEVICES

BASIC ANALYSIS ON DUCTLESS SMALLER DEVICE

A smaller than usual split framework, is both a cooling and warming framework in one that capabilities utilizing heat siphon innovation. By introducing a solitary outside condenser intended to scatter heat, it is feasible to control various indoor evaporator frameworks that can cool or intensity each room at free temperatures. Dissimilar to other customary focal cooling frameworks, you can introduce a ductless split A/C with practically no pipes, yet rather through a line set that goes through an opening in the wall to convey cold refrigerant inside and hot refrigerant outside.

HOW TO FIX A DRYER

Hold board

To eliminate the back board, eliminate a progression of screws or fasteners that hold the board to the top and sides of the bureau. To eliminate the lower front board, pull it away from the lower part of the bureau. Lift the lower board far up into the clouds. Eliminate the springs under each side of the lower board.

Wedge point

To eliminate the top board, wedge a solid bladed clay blade under the edge of the top and pry the top off. The clay blade helps discharge a few spring latches at the highest point of the bureau sides and front. To eliminate the dryer's control board, eliminate the screws that hold it to the bureau top or front.

Screw fine

These screws might be under a piece of metal or plastic trim; pry off or unscrew the trim. Most handles are grating fit; pull them straight out. While testing or fixing the electrical pieces of a gas dryer, recall that the dryer is snared to a gas pipe.

Off available

Switch off the shutoff valve on the stockpile pipe prior to disengaging the gas supply line or moving the dryer and prior to accomplishing any electrical work. Other than drying garments, dryers likewise eliminate build up. This fine, fluffy material can create problems since it blocks dryer build up traps, obstructs vents, and fills blowers. Build up can likewise assemble around and in the tracks of the drum rollers, or in and under the pulleys and the drive belt.

Massive build

The outcome is unfortunate garments drying or at times no drying. To keep away from buildup issues, clear out the dryer's build up trap framework each time you utilize the dryer. To clean the buildup screen, eliminate it from the unit. The screen might be situated close or under the entryway ledge, or in the highest point of the dryer close to the control board. It can normally be eliminated by pulling it up and out of its lodging.

Clean level

Eliminate the amassed build up to clear the screen; then, at that point, supplant the screen. The exhaust vent additionally gathers build up. Vent upkeep includes cleaning the build up from a screen in the dryer's vent exhaust collar or potentially toward the finish of the exhaust vent where it stands out through a cellar window or through an outside wall.

Brush vent

To clean the screen, eliminate the brace that holds the vent to the collar, or back out the screws that hold the vent to the collar, or pull the vent straight off a lengthy collar. Clean the screen completely and supplant it in the vent get together.

Line or tube

To clean the actual vent, utilize a vent-cleaning brush accessible at home improvement shops to take out any build up stores. Likewise check the vent hurry to ensure that the vent channeling or tubing isn't free at the joints, on account of adaptable plastic venting isn't drooping between hanging sections. Breaks or hangs because excessive stress on the dryer's blower framework, and can cause drying issues.

Track fill

On the off chance that the vent line or tubing has become obstructed with buildup, eliminate the build up by pushing a nursery hose or a channel and trap drill through the vent to a helpful joint. Dismantle the joint to eliminate the trash. With this methodology, it isn't important to dismantle the whole vent to track down the blockage.

GUIDE ON USING INDOOR REGULATOR

The Power

To check the control board indoor regulator, ensure power is set for the dryer. Test the indoor regulator with a VOM set to the RX1 scale. Cut one test of the VOM to each indoor regulator terminal. On the off chance that the meter understands zero, the indoor regulator is working. On the off chance that the needle leaps to a high perusing, the indoor regulator is defective and ought to be supplanted. If vital, supplant the indoor regulator with another one of a similar sort.

Verify

Interface the new indoor regulator the same way the bygone one was associated. To check a working indoor regulator, first verify that the capacity to the dryer has been switched off and the dryer is cool. Then detach the prompts one side of the indoor regulator.

Meter

Test a working indoor regulator with a VOM set to the RX1 scale; cut one test to every terminal of the indoor regulator. Separate the prompts one side of the indoor regulator with the goal that the meter won't give a bogus perusing. In the event that the meter understands zero, the indoor regulator is working. In the event that the needle leaps to a high perusing, the indoor regulator is flawed and ought to be supplanted.

STEPS-BY-STEPS ON HVACS

Materials

Drill

Opening saw

Customizable or force wrench

Level

Condenser wall sections

Ductless small parted framework

A plastic condenser cushion for outside unit to lay on you can likewise pour your own substantial piece to make a level surface

Plastic line set cover pack to contain all lines

INSTRUCTIONS

Size selections

Before you start looking for your ductless split A/C, you want to verify you pick the right size for your space. Which alludes to size cooling or warming limit of the framework; The higher the more noteworthy the limit; to begin, you should compute the area of your room so you can pick a little parted framework that is reasonably for that space.

Units place

Whether you select yours in light of the size of the room, you should pick an accurately measured ductless split A/C. One that is too little won't cool adequately, and one that is too huge will turn on and off regularly to keep up with the temperature bringing about superfluous mileage. When you have the right unit, you can move onto your smaller than normal split establishment.

Electrical

Start by choosing a wall in the room that you intend to warm or cool and ensure this wall prompts the outside to consider the refrigerant and electrical cables to associate through. Ensure the wall you select has adequate room to hold the air overseer and is likewise liberated from any impediments, for example, retires or lights.

Lines air up

The air overseer ought to quantify at least 6 feet from the floor to the lower part of the unit. Make certain to observe producer's rules for the rooftop leeway as well. Guarantee that the outside wall is additionally liberated from impediments as you will run lines from the inside to the open air unit through this wall.

Mount

It's ideal to introduce the indoor air overseer plate straightforwardly onto the wall studs, so try to find them with a stud locater in the mounting area.

Make specific the mounting opening areas on the mounting section line up with the stud design. You will likewise be boring a 4 inch opening through the wall to the outside, so ensure the opening area evades the stud.

Sections roll

Moreover, you really want to guarantee there are no lines, pipes, wires, or whatever else in the wall that might cause an impediment. You can then put the section for holding the wall unit against the wall and utilize a level to accurately adjust it. On the off chance that the air overseer isn't level, the channel may not work as expected. Mark those mounting focuses onto the wall and begin by boring pilot openings with a bore that is somewhat more modest than the shaft of your expected screws.

Drill

Drill a pilot opening in the wall for the line set refrigerant lines, channel tubing, and control lines. You can utilize the mounting section to stamp this opening commonly on the lower left-or right-hand corner. The width of the opening is subject to the lines gave however is for the most part around 1 - 3 inches. Some air controllers provide you with the decision of which side the lines will leave the unit yet use makers directions to figure out how to move them assuming need be.

Adjust after drill

Drill an opening the size of the opening saw's pilot bit, sufficiently long to leave the outside of the house at a somewhat descending point so you will actually want to adjust the bigger opening you drill with the opening saw. Then, at that point, utilize the opening saw spot to bore the 2 or 5 inch opening, calculating it descending somewhat for better waste. Line set openings should incline somewhat descending to the outside so the condensate can deplete outside. Really look at the outside to guarantee there are no deterrents then you can bore a regular opening.

Wires up

Associate the control wires and run the line set to your indoor unit as indicated by maker's directions. Typically this includes the evacuation of the back plate to connect the wire the hard way. Unfurl and push the refrigerant line through the opening and utilize electric tape to keep the cooler line, control wire, and seepage tubing along effortlessly. Raise your indoor unit toward the mount and pull every one of the lines, wires, and tubing through the opening. Then mount the unit safely utilizing gave equipment.

Feed and coil

Uncoil the line set from the rear of evaporator and feed through the opening from outside utilizing an associate here will help. Point the line set descending and to the side to arrive at the condenser. Remember the channel line is most limited of all and goes through last - underneath different lines. Whenever you have introduced your indoor unit, you can move onto the open air condenser. Pick where you intend to introduce your open air condenser unit - whether this is on the ground, rooftop, or on an external wall utilizing mounting equipment.

Condenser

In the event that you put it on the ground, you should tie down it to a condenser cushion which you can either buy or make from concrete. When you introduce the condenser, you can associate the lines, wires, and cylinders from within unit. Securely connect the wires and tubing to the outside wall and utilize a cover for a more smoothed out look.

Transmit and line

Interface the right line joint to the line set on the rear of the outside condenser unit. Then you can interface the multi-transmitter wiring that runs from the indoor to the outside unit. This is where you will require a force wrench.

PRACTICAL REASON OF AC MAKING COMMOTION

Harmed capabilities, like refrigerant harm or spilling. Unfortunate establishment that has prompted broken or removed parts or stuck or broken refrigeration line.

Free parts or potential a missing or broken segregation, Spillage in the refrigerant region which should be tackled as quick as possible; A blower that has broken as well as failed;

WAYS TO LESSEN OUTSIDE A/C CONTROL SYSTEM COMMOTION

Wall or protecting screen

Commotion can return quickly from hard surfaces onto neighbor's property similarly that light is reflected from glossy surfaces. Try not to put the climate control system close to various reflecting surfaces, like walls and roof. Similarly, existing obstructions, for example, fencing and nursery walls, might be used to screen clamor from neighbors effectively. Any snags used to protect neighbors from commotion should impede the line of vision of the air conditioner units and shouldn't have any breaks or holes.

Soundproof nook or a soundproof box

Having an additional boundary that can obstruct and retain the sound can help a ton in diminishing A/C clamor. One exact technique is utilizing an acoustic cover to soundproof an outside A/C unit. Soundproofing covers are perfect for this, especially those explicitly planned to lessen A/C commotion. You might try and build one for your climate control system, like the soundproof boxes utilized for generators and air blowers.

www.ingramcontent.com/pod-product-compliance
Lightning Source LLC
Chambersburg PA
CBHW070122010626
45794CB00012B/1216